LAYERS OF THE EARTH

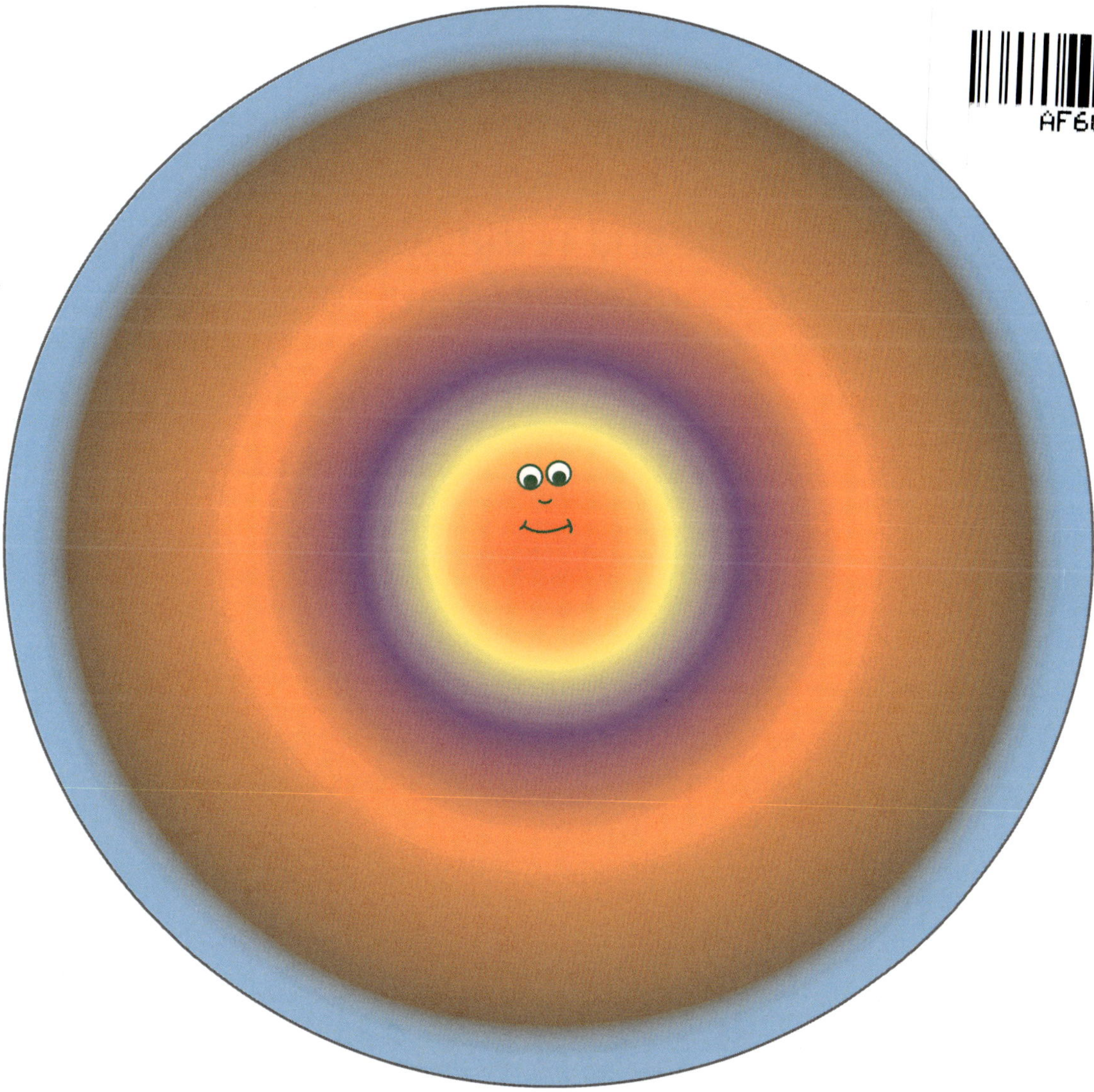

KlevaKids

Published by KlevaKids.com Inc

New York, New York

For more titles,visit our website at www.klevakids.com

KlevaKids

KlevaKids

Presents

LAYERS OF THE EARTH

The earth's layers are introduced in the context of a story about the formation of the earth.

The above statements are purely informational. The purpose of the book is not to teach, but to expose concepts. Read the book to a child and just allow the story to be enjoyed.

It was a gas cloud

that was quite amazing to see.

Layers of the earth formed from it.

That's why this cloud is key.

Inside the cloud

there was lots of activity.

Substances were pulled to the center

by an invisible force called gravity.

The first substances included

lots of iron ore.

They became solid under pressure

and were called the inner core.

Around the inner core,

the outer core started to settle.

It was liquid and hot,

called molten metal.

Separating the inner and outer cores

formed the Inge Lehmann Discontinuity.

Inge Lehmann discovered that boundary.

She was a seismologist with ingenuity.

The movement in the outer core

is often called kinetic.

It spins and generates a field

that is known as magnetic.

The lower mantle then formed

around the outer core.

Gravity pulled it in so tight

that it seemed like such a chore.

Between the outer core and lower mantle

is the Beno Gutenberg Discontinuity line.

Beno Gutenberg studied geophysics

and that was mighty fine.

The lower mantle is solid rock,

but not like the kind you throw.

There is so much heat and pressure

that the rocks move like rivers going slow.

The upper mantle also has rock

and amazingly it too can flow.

It does not move fast.

In fact it moves real slow.

It started to wrap

the upper mantle so gently and neatly.

It was the earth's crust

that would cover the mantle completely.

The Moho Discontinuity

separates the mantle and the crust.

It was named after Andrija Mohorovicic,

a seismologist we came to trust.

The earth's crust is solid

and we know that very well.

We see it all around us.

It is the place where humans dwell.

When you see the continents and oceans you can beam with a lot of pride. You know that the earth has layers deep down inside.

THE END

GLOSSARY

Geophysics: the study of how the earth works

Seismologist: a person who studies earth quakes and other vibrations of the earth

EARTH CROSS SECTION

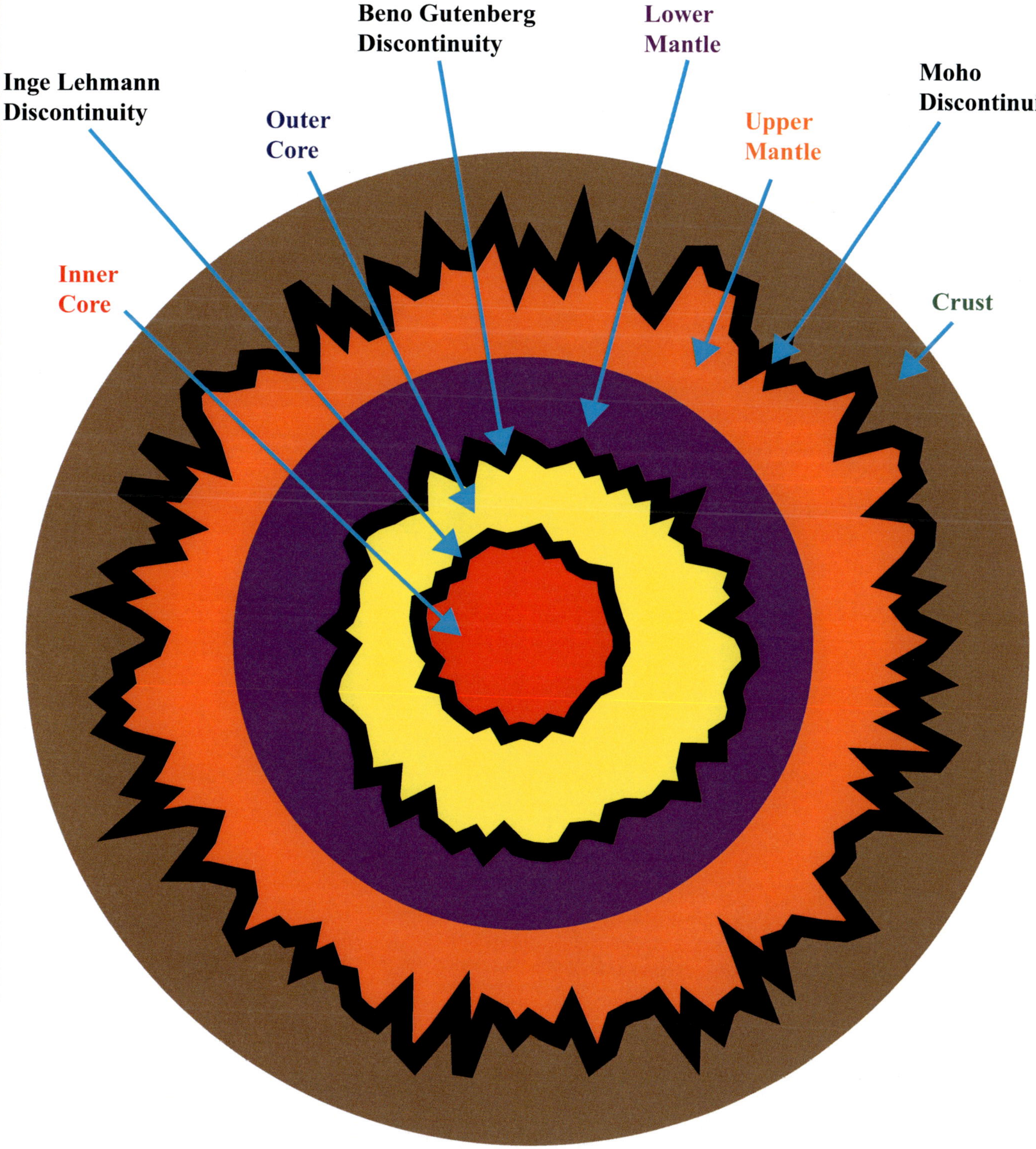

39309024R00021

Made in the USA
Middletown, DE
11 January 2017